The National Atomic Museum

America's Museum Resource for Nuclear Science and History

Sam Bono

W9-CBD-464

THE
DONNING COMPANY
PUBLISHERS

The Donning Company Publishers
184 Business Park Drive, Suite 206
Virginia Beach, VA 23462

Steve Mull, General Manager
Ed Williams, Project Director
Pam Forrester, Project Research Coordinator
Dawn V. Kofroth, Assistant General Manager
Sally Clarke Davis, Editor
Marshall McClure, Senior Graphic Designer
John Harrell, Imaging Artist
Scott Rule, Director of Marketing
Patricia Peterson, Marketing Coordinator

Library of Congress Cataloging-in-Publication Data

Bono, Sam, 1947–
      The National Atomic Museum : America's museum resource for nuclear
   science & history / written by Sam Bono.
         p. cm.
      ISBN 1–57864–146–2 (pbk.:alk. paper)
      1. National Atomic Museum (U.S.)—History. 2. Atomic bomb—
   Museums—New Mexico—Albuquerque—History. 3. Nuclear physics—
   Museums—New Mexico—Albuquerque—History. I. National Atomic
   Museum (U.S.) II. Title.

   QC773.3U5 B66 2001
   539.7'074'78961—dc21
                                                        2001047010

Printed in the United States of America

On the Cover: Military and civilian VIPs wearing protective goggles watch an atmospheric nuclear test from the Officers' Beach Club patio on Parry Island, Enewetak Atoll, on April 8, 1951. The test, named Greenhouse Dog, was 12.5 miles away on Runit Island. *(Photo courtesy of the USAF Lookout Mountain Laboratory)*

CONTENTS

ACKNOWLEDGMENTS

I am indebted to a number of people at the National Atomic Museum and at Sandia National Laboratories for their help in writing this book. In particular, I would like to thank National Atomic Museum Foundation (NAMF) staff members Shauna Jennings and Tony Sparks. I would also like to thank all those who kindly supplied me with the numerous photos used in this book, and NAMF board member Harold Rarrick who took a number of them.

In the publication of this book, I must thank Ed Williams, the Project Director for the Donning Company Publishers, and Sally Davis, my editor.

INTRODUCTION

The mission of the National Atomic Museum is to serve as America's museum resource for the fascinating history and science of the "Atomic Age." Our goal is to present this important information with engaging exhibits and educational programs that convey the wide diversity of individuals and exciting events that have shaped the historical and technical context of this age. As part of this goal we attempt to present this information in as neutral a format as possible to enable Museum visitors to reach their own conclusions.

The start of the Atomic Age has been linked to different scientific discoveries and events. The most dramatic and the

The planned new atomic museum will be named The National Museum of Nuclear Science and History. It will be located on the north side of Albuquerque, just west of Interstate 25, on the Albuquerque Balloon Fiesta Park, and adjacent to the new Anderson-Abruzzo International Balloon Museum. *(Drawing courtesy of DCSW Architects)*

one with the most potential impact on the future of mankind was the world's first atomic test on the morning of July 16, 1945. The story of this earth-shattering detonation, part of America's successful $2.2 billion-dollar effort to build the first atomic bomb during World War II, has long been presented in the Museum.

While the atomic bomb is only one part of the history and science of the Atomic Age, lack of space has prevented the Museum from telling the entire story. Currently a new National Atomic Museum with a new name and location is being planned. This exciting new museum will enable us to present a more complete history of the Atomic Age.

Unfortunately heightened national security after the tragic events of September 11, 2001, forced the closure of the Museum in its longtime home on Albuquerque's Kirtland Air Force Base and into an interim location. Because of publishing deadlines and since this location is but a stepping stone between our old one of over thirty-two years and a much larger new museum, it is not reflected in this book.

Inset: **Main entrance to the National Atomic Museum sometime in 1978, before an extensive remodeling in 1994 gave the Museum its present appearance.** *(Photo courtesy of Sandia National Laboratories)*

Below: **July 16, 1945, ten seconds after detonation, a gigantic mushroom cloud produced by the world's first atomic test rises thousands of feet above Trinity Site in the New Mexico desert, 230 miles south of Los Alamos where the Trinity device was built.** *(Photo courtesy of Los Alamos National Laboratory)*

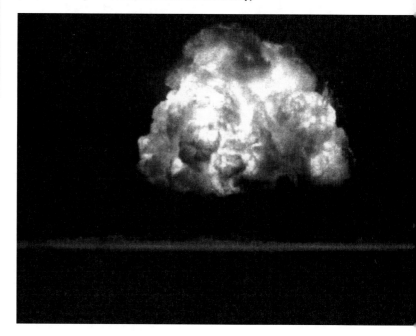

Marie Curie, born
Manya Sklodowska in
Warsaw, Poland, on
November 7, 1867, shared
her first Nobel Prize for
her work in isolating
plutonium with her
French husband, Pierre
Curie. *(Photo courtesy
of the American Institute
of Physics (AIP), Niels
Bohr Library)*

① A Short History of the Atomic Age

BEFORE THE BOMB

This would be far more than a short history of the Atomic Age if we discussed all the history and events that led up to the creation of the first atomic bomb by the United States. Hard work by numerous scientists across the world and throughout the years eventually resulted in the splitting of the uranium atom in December, 1938, by German scientists Otto Hahn and Fritz Strassmann. In doing so, the Germans proved Albert Einstein's famous $E=mc^2$ formula, and that the unknown matter (m) in the formula was natural uranium, U-238. The discovery would lead directly to the building of the atomic bomb by the United States less than seven years later.

The names of these men and women make up a who's who of famous scientists, many of whom were awarded the prestigious Nobel Prize for their pioneering work. Marie Curie, for example, became the first woman to win a Nobel Prize (Physics–1903), and the only woman to win a second one (Chemistry–1911) for isolating the element radium, a vital step in the process toward building the bomb. Marie's daughter, Irene

Above: **The German-born, naturalized American citizen, Albert Einstein, is considered by many to be the greatest theoretical physicist of the twentieth century.** *(Photo courtesy of the Lotte Jacobi Archives Photographic Services, University of New Hampshire)*

7

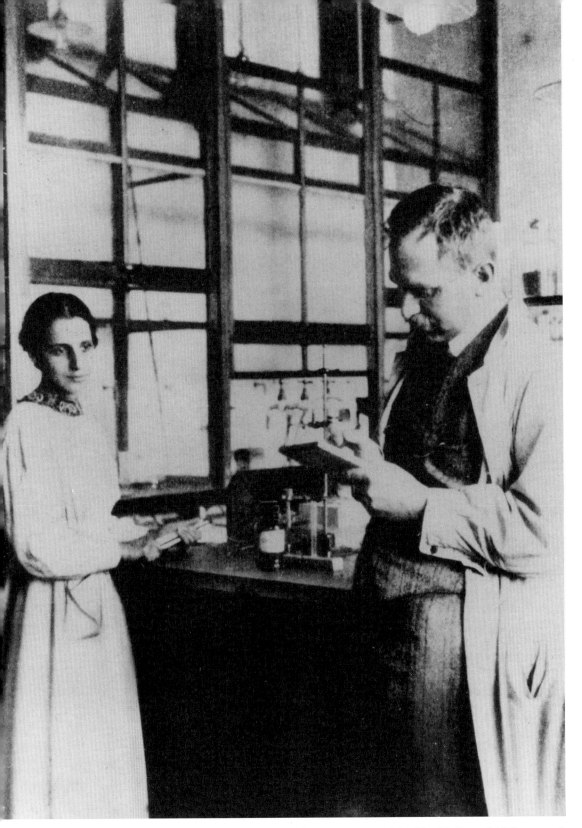

Chemist Otto Hahn with his coworker and friend, Austrian physicist Lise Meitner, laid the ground-work for fission in Germany while working at the Kaiser Wilhelm Institute for Chemistry in the Berlin suburb of Dahlem. *(Photo courtesy of American Institute of Physics (AIP), Niels Bohr Library)*

Joliot-Curie, and her husband, Frederic, won a third Nobel Prize (Chemistry–1935) "in recognition of their synthesis of new radioactive elements."

Other Nobel Prize winners involved in the science of the Atomic Age included: British physicist Ernest Rutherford (Chemistry–1908), who theorized the existence of the atomic nucleus; the German-born Albert Einstein (Physics–1921); and American Ernest Orlando Lawrence, the inventor of the cyclotron and the discoverer of plutonium (Physics–1939). Also heavily involved were Italian Enrico Fermi (Physics–1938), and British physicist James Chadwick (Physics–1935), who discovered the neutron.

Many other important scientists, whose work would lead up to the building of the bomb, either did not win a Nobel Prize or lived and made their great discoveries before the creation of the famous award in 1901.

THE ATOMIC BOMB

In a letter dated December 19, 1938, Otto Hahn wrote and advised his friend and former colleague, Austrian scientist Lise Meitner, of his and Fritz Strassmann's important discovery. Meitner, working with her nephew, Otto Frisch, also a scientist, over the Christmas holidays realized the full impact

Born in Rome on September 29, 1901, Enrico Fermi *(above)* **earned his doctorate from the University of Pisa by age twenty-one. In 1938, Fermi left his native Italy for the United States after winning the Nobel Prize in Physics because of conflicts with the Italian Fascists.** *(Photo courtesy of Los Alamos National Laboratory)*

Ernest Orlando Lawrence was described in his 1968 biography by Herbert Childs as "an American genius," and "the architect of the modern large-scale, team research programs in basic science, exemplified most prominently by the national laboratories." *(Photo courtesy of the Lawrence Berkeley National Laboratory)*

of the Germans' discovery and they named the newly discovered process "fission," a term used in biology to describe the splitting of a cell.

Frisch returned to his job in Copenhagen where he worked with the brilliant Danish theoretical physicist, Niels Bohr, and related the exciting news to Bohr on January 3, 1939. A few days later, Bohr sailed to the United States where he soon revealed to the U.S. scientific community and to the world the Germans' great discovery.

With the help of the world famous Albert Einstein, concerned scientists in the U.S., many of whom had fled from Fascist and Nazi repression in Europe, warned the American government of the possibility of the Germans building an atomic bomb. This warning, in the form of a letter signed by

Above: **After revealing the secret of fission to the participants of the Fifth Washington, D.C. Conference on Theoretical Physics in early February, 1939, Neils Bohr went on to join the Manhattan Engineer District and worked at Los Alamos, New Mexico.** *(Photo courtesy of the Nobel Institute)*

Einstein, eventually would lead to the creation of the Manhattan Engineer District (MED) in August 1942, less than one year after the Japanese attack on Pearl Harbor and America's entry into World War II.

More commonly called the Manhattan Project, the MED was the top-secret scientific-engineering effort created by the United States to build an atomic bomb before the Germans or the Japanese. Under the control of the U.S. Army Corps of Engineers, the Project would involve over 250,000 scientists, engineers, technicians, and many others in three major facilities and in many smaller ones across the U.S.

The efforts of the Manhattan Project would culminate, less than three years after its creation, in the world's first nuclear test at New Mexico's Trinity Site on the morning of July 16, 1945. Two subsequent atomic bombs would be dropped on Japan, bringing World War II officially to an end a month later on August 15, 1945.

The atomic bomb would soon be a major factor in another war: the Cold War. This long, mostly nonshooting

Left: **The discovery that the uranium atom could split (a process soon to be named fission) by chemist Otto Hahn (pictured) and his colleague Fritz Strassmann at Germany's Kaiser Wilhelm Institute, ignited the race to build the first atomic bomb.** *(Photo courtesy of the AIP Neils Bohr Library)*

continued on page 16

Nazi dictator Adolf Hitler (foreground in center at Nuremberg, Germany), fortunately for Europe and the world, never fully understood the potential of the atomic bomb. *(Photo courtesy of the World War II Collection of Seized Enemy Records in the National Archives)*

Army Sgt. Herbert Lehr delivers the plutonium core and neutron initiators for the Trinity Device to the George McDonald Ranch house on July 12, 1945. Note the twenty "rubber corks" mounted on the black magnesium carrying case, that were designed to absorb shocks. *(Photo courtesy of Los Alamos National Laboratory)*

Albert Einstein
Old Grove Rd.
Nassau Point
Peconic, Long Island

August 2nd, 1939

F.D. Roosevelt,
President of the United States,
White House
Washington, D.C.

Sir:

Some recent work by E. Fermi and L. Szilard, which has been communicated to me in manuscript, leads me to expect that the element uranium may be turned into a new and important source of energy in the immediate future. Certain aspects of the situation which has arisen seem to call for watchfulness and, if necessary, quick action on the part of the Administration. I believe therefore that it is my duty to bring to your attention the following facts and recommendations:

In the course of the last four months it has been made probable - through the work of Joliot in France as well as Fermi and Szilard in America - that it may become possible to set up a nuclear chain reaction in a large mass of uranium, by which vast amounts of power and large quantities of new radium-like elements would be generated. Now it appears almost certain that this could be achieved in the immediate future.

This new phenomenon would also lead to the construction of bombs, and it is conceivable - though much less certain - that extremely powerful bombs of a new type may thus be constructed. A single bomb of this type, carried by boat and exploded in a port, might very well destroy the whole port together with some of the surrounding territory. However, such bombs might very well prove to be too heavy for transportation by air.

Dated August 2, 1939, only one month before the start of World War II, the "Einstein Letter" warned U.S. President Franklin D. Roosevelt "that extremely powerful bombs of a new type [atomic bombs] may thus be constructed." (Letter courtesy of the Franklin D. Roosevelt Library)

The United States has only very poor ores of uranium in moderate quantities. There is some good ore in Canada and the former Czechoslovakia, while the most important source of uranium is Belgian Congo.

In view of this situation you may think it desirable to have some permanent contact maintained between the Administration and the group of physicists working on chain reactions in America. One possible way of achieving this might be for you to entrust with this task a person who has your confidence and who could perhaps serve in an inofficial capacity. His task might comprise the following:

a) to approach Government Departments, keep them informed of the further development, and put forward recommendations for Government action, giving particular attention to the problem of securing a supply of uranium ore for the United States;

b) to speed up the experimental work, which is at present being carried on within the limits of the budgets of University laboratories, by providing funds, if such funds be required, through his contacts with private persons who are willing to make contributions for this cause, and perhaps also by obtaining the co-operation of industrial laboratories which have the necessary equipment.

I understand that Germany has actually stopped the sale of uranium from the Czechoslovakian mines which she has taken over. That she should have taken such early action might perhaps be understood on the ground that the son of the German Under-Secretary of State, von Weizsäcker, is attached to the Kaiser-Wilhelm-Institut in Berlin where some of the American work on uranium is now being repeated.

Yours very truly,

A. Einstein

(Albert Einstein)

Above: **The assembled plutonium core for the Trinity Device is loaded into the backseat of a 1942 Plymouth Special DeLuxe for the short trip to the Trinity Site, only two miles north of the McDonald Ranch house on the morning of July 14, 1945.** *(Photo courtesy of Los Alamos National Laboratory)*

After the core was inserted into the round Trinity Device, or "gadget" as it was sometimes called, it was hoisted to the top of a one-hundred-foot steel tower. The large wooden container at left was filled with concrete and used to test the capacity of the tower's winch. *(Photo courtesy of Los Alamos National Laboratory)*

The approximately 20-kiloton (equal to 20,000 tons of TNT) Trinity Device would vaporize the massive steel tower. The two thick coaxial cables extending from the north side of the tower were used to collect data. *(Photo courtesy of Los Alamos National Laboratory)*

war between the United States, the Soviet Union, and their allies lasted from shortly after the end of World War II until the collapse of the Soviet Union in 1989.

THE COLD WAR

Most of the nuclear weapon shapes and delivery systems on exhibit at the National Atomic Museum are a direct result of the almost forty-year Cold War between the western powers led by the United States and the former Soviet Union and her Warsaw Pact Allies. Changes in the world's political situation, large and small "hot" wars, and advancements in nuclear weapon design by one Cold War opponent or the other drove the development of newer and more powerful nuclear weapons by both sides during this "war."

The Cold War started a few years after the end of World War II when for various reasons the Soviet Union, one of America's World War II allies against Nazi Germany, became her opponent in a "cold war." The Cold War took a dramatic and potentially very dangerous turn in late August 1949, when America's four-year monopoly on nuclear weapons was broken by the Soviets. On August 29, 1949, the Soviet Union exploded its first nuclear weapon, which was nick-named "Joe 1" by the U.S. This test took place long before it was expected by the U.S., and helped American President, Harry S. Truman, decide to give the go ahead for the

Jubilant workers at the gigantic Manhattan Project facility in Oak Ridge, Tennessee celebrate the end of War World II. *(Photo courtesy of Oak Ridge National Laboratory)*

Scientists stand in front of "Mike," the first large-scale thermonuclear device (equal to 10.4 million tons of TNT) on Elugelab Island, Enewetak Atoll. After the October 31, 1952 test, Elugelab disappeared, replaced by a crater 4,500 feet in diameter and over 200 feet deep. *(Photo courtesy of Los Alamos National Laboratory)*

development of the controversial and much more destructive "super" or thermonuclear bomb, which only increased Cold War tensions.

During the years of the Cold War, the U.S. was involved in two large conflicts, Korea (1950–1953) and Vietnam (1965–1973), and in numerous smaller ones, but none involved a direct confrontation with the Soviet Union, or, fortunately, the use of nuclear weapons. According to history, the closest the U.S. and the former Soviet Union came to a fatal nuclear confrontation was during the October 1962 Cuban Missile Crisis (known in the Soviet Union as the Caribbean Crisis).

An unfortunate consequence of U.S. Cold War nuclear preparedness has been the thirty-two reported nuclear weapons accidents in the thirty years between 1950 and 1980. The best known of these so-called "Broken Arrows" is

Above: **U.S. President Harry S. Truman, having made the decision to use the atomic bomb to help end World War II, was not willing to make the same decision during the Korean War six years later.** *(Photo courtesy of the U.S. Army and the Harry S. Truman Presidential Library)*

Left: **Eighty days after the January 17, 1966, Palomares, Spain, nuclear accident, the Submarine Rescue Vessel, USS *Petrel* (ASR–14), hauled the last weapon involved out of the Mediterranean Sea. The Mk 28 bomb was recovered from a depth of 2,800 feet.** *(Photo courtesy of the U.S. Air Force)*

BM LAUNCH SITE 3
SAN CRISTOBAL, CUBA
27 OCTOBER 1962

LAUNCH AREA

NUCLEAR WARHEAD BUNKER U/C

PERMANENT BLDGS

STORAGE

TRENCH

This American surveillance photo shows a nuclear-capable
Medium Range Ballistic Missile (MRBM) launch site in eastern
Cuba. The Soviet SS-4 MRBM's had a range between 600 and 1,500
miles and could target most of the eastern United States. *(Photo
courtesy of the John Fitzgerald Kennedy Library)*

Left: **Descended from the Army's Nike-Ajax, Nike-Hercules, and the Nike-Zeus A, the three-stage Nike-Zeus B was designed to intercept targets outside the earth's atmosphere. Like many early missile systems, it was tested at New Mexico's White Sands Missile Range.** *(Photo courtesy of the National Archives)*

the one involving four nuclear weapons that took place over Palomares, Spain, in January 1966. (Learn all about the Palomares accident and view the two heavily damaged surviving nuclear bombs in the Museum.)

This ongoing development, testing, and construction of nuclear weapon systems and their related delivery systems during the Cold War has been estimated to cost the United States somewhere between 3 and 4,000,000,000,000 (trillion) dollars. In the Soviet Union, the enormous cost of nuclear weapon development was one of the causes that eventually forced the breakup of the USSR into fifteen separate republics in December 1991.

NUCLEAR POWER

The uncontrolled fission of uranium 235 or plutonium (Pu 239) in a bomb can result in a nuclear explosion. These

The long range and pinpoint accuracy of the Pershing II convinced the Soviet Union to seek a treaty (the Treaty on Intermediate Range Nuclear Forces) with the U.S. that eliminated a variety of nuclear-capable weapons, including the deadly Pershing.
(Photo courtesy of the National Archives)

Above: **This August 11, 1958 photo shows a Navy Regulus II missile aboard the USS *Grayback* (SSG–574), the first submarine to successfully fire it. While a success, the Regulus II was soon canceled in favor of the more powerful and longer range Polaris missiles.** *(Photo courtesy of the U.S. Navy)*

same materials controlled and when used in a nuclear reactor can produce a tremendous amount of electrical power.

On December 20, 1951, a nuclear reactor located in the desert of eastern Idaho was the first to produce electricity using atomic energy. Named the Experimental Breeder Reactor-I, or EBR-I, this reactor is located at what is now the Idaho National Engineering and Environmental Laboratory, fifty miles west of Idaho Falls, Idaho. (Now a National Historic Landmark, EBR-I is open to the public). While EBR-I initially produced only enough electrical power to light four light bulbs, it proved that the concept was practical. Today, over thirty countries around the world use some 440 commercial nuclear reactors to help provide their electrical needs. In eighteen of these countries, nuclear reactors provide 25 percent or more of their electricity. In France, over 75 percent of its electrical needs are produced with nuclear reactors.

How do nuclear reactors work? The splitting (fission) of uranium atoms in a reactor produces heat, which is used to turn water into steam. The steam is then used to spin an electrical turbine, producing electricity. While simple in concept, nuclear power can potentially be very dangerous

Because of the shortage of uranium reactor fuel in the late 1940s, EBR-I was designed as a breeder nuclear reactor, in that it could produce more fuel than it actually used. *(Photo courtesy of the Idaho National Engineering & Environmental Laboratory)*

This photo of the major components, subassemblies, and test equipment of a B 83 nuclear bomb gives an idea of the complexity of modern U.S. nuclear weapons. The B 83 is 18 inches in diameter, 12 feet long, and weighs 2,400 pounds. *(Photo courtesy of Sandia Livermore National Laboratory)*

The United States has 104 operating nuclear power plants located in thirty states. The majority of these plants are east of the Mississippi. Another twenty-four U.S. nuclear power plants are no longer in operation. *(Photo courtesy of the Nuclear Energy Institute)*

Spent reactor fuel rods are lowered into a water-filled storage pool (usually adjacent to the reactor) until a long-term storage facility is a reality. *(Photo courtesy of the Nuclear Energy Institute)*

if it is not properly controlled.

Of course, the production of electricity by many non-nuclear methods, including coal, oil, and water (hydroelectric) all have associated problems, both potential and actual. These hazards include various forms of pollution (pre and post use), and many dangerous production conditions. Nuclear power's primary unique hazard is radiation.

Since the uranium fuel rods used to power nuclear reactors are radioactive, they must be safely handled and disposed of. The safe storage and disposal of the spent, but still radioactive nuclear fuel rods, is a major concern of the nuclear power industry. In many countries, this spent fuel is reprocessed and reused, while in the U.S. much of it is stored, most often in water filled pools near the reactors, until more permanent storage solutions become available.

While we have the ability to build safe nuclear reactors, they are expensive and must always be properly maintained and operated by a highly trained staff.

Despite the potential problems and cost of nuclear energy production, many countries that do not have oil or coal resources or the geography and water required for hydroelectric power find nuclear power a viable alternative.

NUCLEAR MEDICINE

Some of the same equipment that produced the enriched uranium for the Little Boy atomic bomb at the Manhattan Project facility in Oak Ridge, Tennessee, was later used in 1946 to produce nuclear medicines.

Nuclear medicine is the use of a variety of radioactive materials/radioisotopes to help diagnose or treat different diseases. While not as well known as other medical diagnostic procedures, over forty 40,000 nuclear medicine prescriptions are written daily in the U.S. alone! The majority

Right: **Scientist Hal Anger built the first "scintillation [gamma] camera" used in nuclear medicine tests in 1958.** *(Photo courtesy of the Lawrence Berkeley National Laboratory)*

Below: **The gigantic 2,000,000-square-foot K-25 Gaseous Diffusion Plant that helped produce enriched uranium 235 for the Little Boy atomic bomb was used after the war by the Atomic Energy Commission (AEC) to produce some of the first nuclear medicines.** *(Photo courtesy of Oak Ridge National Laboratory)*

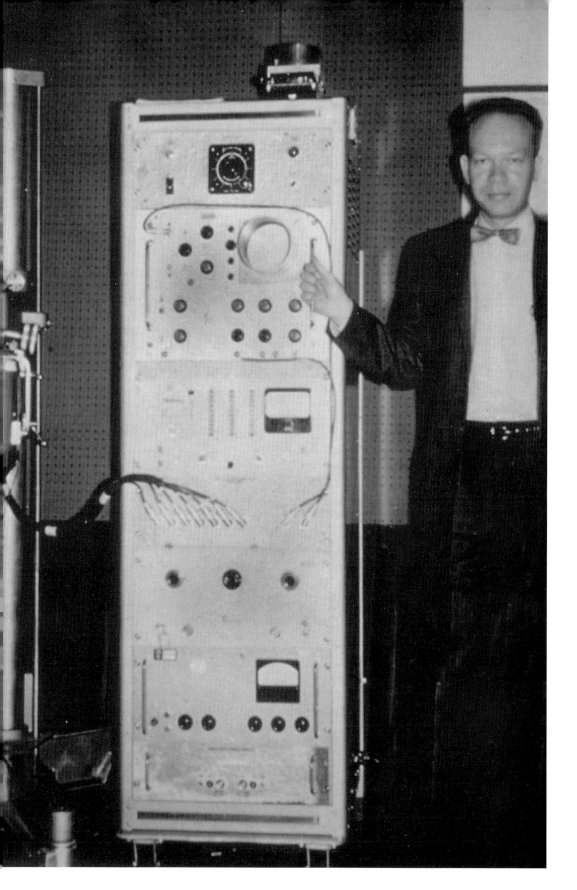

of these daily prescriptions call for low levels of radioactive materials for diagnostic purposes, but the small percentage for treatment involve high levels of radiation. In addition, radioactive isotopes are used in over 100,000,000 medical laboratory tests each year.

Nuclear medicine is so popular because it can diagnose certain diseases, such as heart and thyroid problems, earlier than most other methods, possibly making their treatment more effective. It also has the advantage over surgery of being noninvasive. Different nuclear medicines are also used for blood flow and lung ventilation studies, plus brain, bone, liver, spleen, and kidney imaging.

While the technology involved in nuclear medicine is highly complex, the "how" of nuclear medicine can be broken down into a couple of basic steps. First, the patient is given a radioactive compound (a radioisotope most often attached to a chemical compound) which is introduced into the patient's body by injection, swallowing, or inhalation. Different compounds are used to study different parts of the body.

Carried by the chemical compound, the radioisotope travels through the patient's body to the area to be diagnosed. Next, a gamma camera is used to record the gamma radiation given off by the radioisotopes as they collect in the

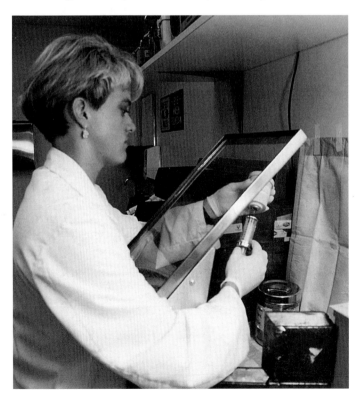

While most nuclear medicines have very low levels of radioactivity, nuclear-pharmacists still take many precautions such as working behind heavily leaded glass shields and lead bricks since they work with the radioactive medicines almost daily. *(Photo courtesy of Sandia National Laboratories)*

"Seeing is Healing" the National Atomic Museum's exhibit on nuclear medicine details the history of this very important medical diagnostic tool. The informative exhibit features a massive gamma camera (above right center) donated to the Museum by Picker International, Inc., of Cleveland, Ohio. *(Photo courtesy of Harold Rarrick)*

area under study. This information is used to create images of that part of the body. Usually, a computer is used to enhance these images to make them easier for the doctor to interpret and then subsequently make an accurate diagnosis.

The results of a nuclear medicine procedure may be combined with results from other diagnostic tests such as x-rays or a MRI to obtain the best analysis of the condition.

Is nuclear medicine safe? Yes, because usually very small quantities of radioactive materials are used, and these materials have half lives measured in only hours or days and are quickly eliminated by the body. Plus, all phases of nuclear medicine procedures are carefully controlled by the nuclear pharmacist who mixes the radioactive compound (usually early on the morning of the test) and by the nuclear medicine technologist who prepares the patient, administers the radioactive compound, and operates the specialized equipment.

As mentioned earlier, much larger doses of radioisotopes are used to help treat some types of health problems such as overactive thyroid glands and some types of thyroid tumors. In addition, powerful radioactive "implants" can be used to shrink or destroy other kinds of tumors. For more information

on the history and use of nuclear medicine, please visit
the Museum's exhibit "Seeing is Healing."

Radioisotopes are also used in food irradiation,
especially in developing countries that lack the transportation and refrigeration facilities found in the United States
and many other countries. Other uses of radioisotopes
include archaeology, environmental science, industrial
applications, criminal investigations, and the sterilization
of medical equipment. Radioisotope nuclear generators
were used to provide electrical power on the moon by the
Apollo astronauts!

NUCLEAR WASTE

One of the more perplexing problems facing the government
and the nuclear industry is the proper disposal of radioactive
waste. Over the last half-century, weapon design, production, and testing have produced large amounts of highly
radioactive nuclear waste. Commercial nuclear power production, nuclear medicine and other uses have also resulted
in large quantities of nuclear waste that has to be safely
stored for a very long time.

The United States Department of Energy (DOE) has proposed two facilities for the long-term storage of this nuclear
waste. The first is the Waste Isolation Pilot Plant (WIPP),
located twenty-six miles southeast of Carlsbad, New Mexico.
The second is inside Yucca Mountain, about one hundred
miles northwest of Las Vegas, Nevada.

**A truck loaded with three
TRUPACT-II containers
filled with fifty-five-gallon
drums (up to fourteen each)
of low-level radioactive
waste pulls up to the Waste
Isolation Pilot Plant (WIPP)
in southeastern New
Mexico.** *(Photo courtesy of
WIPP)*

A part of the 869-ton, 460-foot-long Tunnel-Boring Machine (TBM) protrudes out of Yucca Mountain, the Department of Energy's (DOE) proposed high-level radioactive waste repository northwest of Las Vegas, Nevada. *(Photo courtesy of the U.S. Department of Energy)*

After more than forty years of controversy, WIPP finally opened and received its first shipment of nuclear waste on March 26, 1999. This waste primarily consists of clothing, tools, and other weapon-manufacturing debris contaminated with radioactive elements such as plutonium.

Weapon production waste first began accumulating in the U.S. during the years of the Manhattan Project, and was continually "produced" during the Cold War that followed the end of World War II. With very long radioactive half-lives, this waste will remain dangerous for thousands of years and must be properly stored to protect the environment and future generations. For example, half of the radioactivity in plutonium (Pu 239) that is made today will still be active/dangerous 24,100 years from now!

WIPP dates back to 1956 when the National Academy of Sciences first recommended to the Atomic Energy Commission (the AEC was the successor to the Manhattan Project and the predecessor of ERDA and the DOE) that radioactive waste be disposed of in stable geologic formations, such as deep salt beds. The presence of the thick salt beds prove the absence of circulating ground water that could move waste to the surface (water, if it had been or were present, would have dissolved the salt beds), and salt is relatively easy to mine. Also, salt has the ability to "creep," and will eventually fill a void, thus "sealing" the nuclear repository.

The first choice for the proposed repository was a salt mine near Lyons, Kansas. When this site was rejected by the AEC for various reasons in 1972, a site in the Permian Basin salt deposits of southeastern New Mexico was selected in 1974. In 1979, Congress authorized WIPP, as it was named, and construction began two years later. WIPP was completed in April 1989.

The Waste Isolation Pilot Plant is located in the nearly 3,000-foot-thick Salado Formation, 2,154 feet under the New Mexico desert. These salt beds have essentially remained stable and unaffected by earthquakes or faulting since the ancient Permian Sea deposited them over 225 million years ago.

Between the completion of WIPP in 1989 and its opening ten years later, WIPP, many different U.S. Government agencies, and the State of New Mexico have had to comply with hundreds of applicable environmental regulations. These agencies were also faced with numerous lawsuits from various groups, including one involving the State of New Mexico and three New Mexico environmental organizations against the Environmental Protection Agency (EPA) for certifying that WIPP could safely isolate its waste for 10,000 years.

Now that WIPP is finally open and slowly being filled with waste, it is expected that there is enough radioactive waste currently in storage (at ten different surface locations across the country) or being produced to keep WIPP open until the year 2025 or longer.

Nevada's Yucca Mountain, the proposed second DOE waste storage facility (its proper name is the Yucca Mountain Site Characterization Project), has been officially studied since 1987 (unofficially since 1977) at a cost fast approaching three billion dollars. While WIPP is designed to contain mostly low-level and some medium-level radioactive waste, Nevada's Yucca Mountain will be built to contain and isolate high-level nuclear waste, also the byproduct of weapon development and of the nuclear power industry.

Unlike WIPP which is located in salt beds, Yucca Mountain is composed of welded-tuff, a densely compacted volcanic ash. As at WIPP, one of the main concerns at Yucca Mountain is water. Water can seep down from the surface, or the water table can rise and corrode the waste storage containers and then help spread radiation. Currently the area receives only six to seven inches of rain and snow per year, and the water table is 800 to 1,000 feet below the lowest

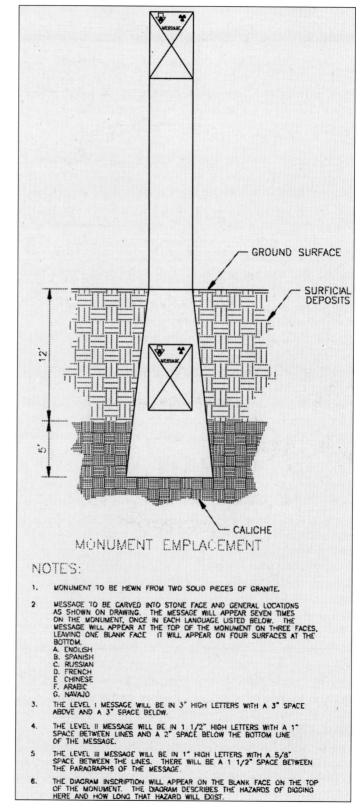

GROUND SURFACE

SURFICIAL
DEPOSITS

12'

5'

CALICHE

MONUMENT EMPLACEMENT

NOTES:

1. MONUMENT TO BE HEWN FROM TWO SOLID PIECES OF GRANITE.

2   MESSAGE TO BE CARVED INTO STONE FACE AND GENERAL LOCATIONS
    AS SHOWN ON DRAWING. THE MESSAGE WILL APPEAR SEVEN TIMES
    ON THE MONUMENT, ONCE IN EACH LANGUAGE LISTED BELOW. THE
    MESSAGE WILL APPEAR AT THE TOP OF THE MONUMENT ON THREE FACES,
    LEAVING ONE BLANK FACE   IT WILL APPEAR ON FOUR SURFACES AT THE
    BOTTOM.
    A. ENGLISH
    B. SPANISH
    C. RUSSIAN
    D. FRENCH
    E CHINESE
    F. ARABIC
    G. NAVAJO

3. THE LEVEL I MESSAGE WILL BE IN 3" HIGH LETTERS WITH A 3" SPACE
    ABOVE AND A 3" SPACE BELOW.

4. THE LEVEL II MESSAGE WILL BE IN 1 1/2" HIGH LETTERS WITH A 1"
    SPACE BETWEEN LINES AND A 2" SPACE BELOW THE BOTTOM LINE
    OF THE MESSAGE.

5   THE LEVEL III MESSAGE WILL BE IN 1" HIGH LETTERS WITH A 5/8"
    SPACE BETWEEN THE LINES. THERE WILL BE A 1 1/2" SPACE BETWEEN
    THE PARAGRAPHS OF THE MESSAGE.

6. THE DIAGRAM INSCRIPTION WILL APPEAR ON THE BLANK FACE ON THE TOP
    OF THE MONUMENT. THE DIAGRAM DESCRIBES THE HAZARDS OF DIGGING
    HERE AND HOW LONG THAT HAZARD WILL EXIST.

Drawing of a proposed WIPP Perimeter Monument. The massive monument will have a warning message carved into three sides of the top and on all four sides of the base in seven languages: Arabic, Chinese, English, French, Navajo, Russian, and Spanish. *(Drawing courtesy of Sandia National Laboratories)*

proposed storage tunnel, but Yucca Mountain will need to be safe for approximately 10,000 years!

Both WIPP and Yucca Mountain, or alternative high-level waste sites, will face another problem when they are filled and closed. How do you mark the site for future generations (remember we are talking 10,000 plus years into the future) so that our ancestors do not accidentally dig, drill, or otherwise open the sites?

WIPP, working with Sandia National Laboratories and groups of experts including historians, linguists, science-related writers, and anthropologists have created what they call Passive Institutional Controls (PICs) to mark the New Mexico site.

WIPP's primary PICs are to include a large earthen berm (with a 100-foot wide base) surrounding the site on the surface, two 25-foot high, 20-ton perimeter monuments, and an "information center" located in the center of the bermed area. Other PICs include two additional sealed information centers/storage rooms, one buried twenty feet underground and one in the berm in addition to randomly buried warning markers made out of different long-lasting materials. Finally, WIPP archives will be stored around the world in different locations.

In addition to the nuclear waste resulting from weapons and nuclear power production, waste, mostly low-level, is generated by many other sources. According to the U.S. Nuclear Regulatory Commission, 858,677 cubic feet of commercially generated radioactive waste was produced in 1994. Nuclear reactors contributed 44 percent, 39.9 percent was from industrial users, 13.4 percent was from government sources (other than nuclear weapons sites), 2.1 percent was from academic users, and 0.6 percent was from medical facilities.

Currently, there are two low-level disposal facilities that accept a broad range of low-level waste. They are located in Barnwell, South Carolina, and Richland, Washington. Four former low-level radioactive waste disposal sites are closed and no longer accept waste. They are located in or near Sheffield, Illinois; Morehead, Kentucky; Beatty, Nevada; and West Valley, New York. The low-level wastes at these six sites are buried under several feet of soil in trenches, usually in the containers in which they were originally shipped.

STOCKPILE STEWARDSHIP

In 1992, U.S. President George Bush signed a moratorium on nuclear tests. This moratorium has been continued by all U.S. presidents since then. As a result, there has been a growing concern about the reliability of America's nuclear

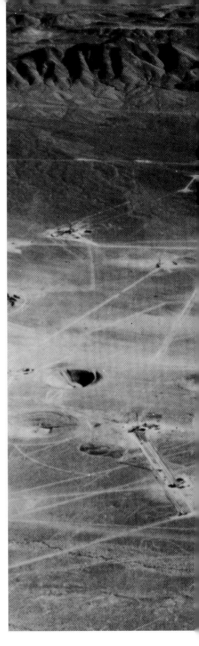

America's last nuclear weapons test, equal in power to the 20 kt Trinity test, was underground at the Nevada Test Site (NTS) on September 23, 1992. This photo shows some of the numerous nuclear test "subsidence craters" that dot the NTS landscape.
(Photo courtesy of the Nevada Test Site)

weapons and whether our Science Based Stockpile Stewardship (SBSS) Program will ensure a viable nuclear deterrent for the future.

The SBSS Program, primarily using computer simulations, has been used since the 1992 moratorium to evaluate the thousands of nuclear warheads in the U.S. stockpile. SBSS Program officials are confident that America's stockpile is safe and secure, and that the program can fully maintain the nation's nuclear weapons without actually detonating a nuclear weapon.

Detractors of the program have concerns about its underlying technical rationale and fear that the program's base of talented scientists is eroding. Other weapon experts believe that the concept of trying to assess the stockpile in the absence of nuclear tests is intrinsically flawed. Without exploding a sample of the bombs, critics say, the United States can never reach a clear-cut conclusion about the continued reliability of any given weapon design.

NUCLEAR TIMEKEEPING

The U.S. Congress established the National Bureau of Standards (NBS) in 1901 to support industry, commerce, scientific institutions, and all branches of government. Eighty-seven years later, in 1988, Congress expanded the mission of the NBS and changed its name to the National Institute of Standards and Technology (NIST) to reflect its new mission.

In the late 1940s, the NBS/NIST became involved in the use of "atomic clocks" for its mission to provide very accurate time to the nation. Using a method suggested by Manhattan Project veteran Isidor Rabi in 1945, NBS created the world's first atomic clock in 1949. This original clock, which used the ammonia molecule as a source of vibrations, was replaced in 1952 by a more accurate clock that was named NBS-1. This clock used cesium atoms as its vibration source.

Since then, NBS/NIST nuclear clocks have improved in accuracy and reliability from NBS-1 through the 1993 NIST-7. The current atomic clock, the NIST-F1 Cesium Fountain Clock, has an accuracy of plus or minus one second in 20 million years!

In our increasingly technical world, the high-accuracy timekeeping provided by NIST-F1 is critical to a number of important systems. For example, modern satellite telecommunications systems require synchronization to better than 100 billionths of a second, and the Defense Department's Global Positioning System (GPS), now in everyday use across the world, is dependent upon the extremely accurate time provided by NIST's atomic clocks.

Accurate time is also important in the ordering of many human activities including the world's financial markets, such as New

Below: **Dr. Edward U. Condon (left) and Dr. Harold Lyons, both of the National Bureau of Standards (now the National Institute of Standards and Technology), stand beside their creation, the world's first atomic clock in 1949.** *(Photo courtesy of the National Institute of Standards and Technology)*

The Kiwi-B4-A nuclear rocket engine was tested on November 30, 1962, at the Nuclear Rocket Development Station (NRDS), Nevada. During the 260-second test, Kiwi reached 50 percent of its design power. *(Photo courtesy of Los Alamos National Laboratory)*

York City's Wall Street Stock Exchange. Time/date stamps are used in stock exchanges worldwide to identify transactions so that they can be placed in order, a process that is becoming increasingly important as electronic commerce moves at faster and faster speeds.

NUCLEAR SPACE POWER

With the planned completion of the earth-orbiting International Space Station in 2005, interest has been growing in the next big leap in space—a possible manned mission to Mars. Currently, the fastest and, therefore, the safest way to get to the red planet would be with nuclear propulsion. Faster is safer in space travel because of the large amounts of radiation from galactic cosmic rays to which astronauts are exposed. So the possibility of making a round

Bimodal Nuclear Thermal Rocket Propulsion

Blueprint for 21st Century Space Travel

trip from the Earth to Mars in less than a year (with a two-month layover on Mars) as compared to a trip powered by conventional chemical rockets, that may take three times as long, is highly desirable. But, this is not the first time the U.S. has been interested in nuclear-powered space travel.

From 1959 until 1972, when the original nuclear space power program was cancelled, the U.S. spent nearly $1.4 billion designing, building, and static testing nuclear-reactor rocket engines intended for deep space travel. With a renewed interest in space travel, a corresponding interest in nuclear rockets, especially the gas-core nuclear rocket (GCNR), has taken place.

Just like conventional chemical rockets, nuclear rockets also need a fuel, but since nuclear engines are much more efficient they can also provide the electricity needed to run the spacecraft, provide for an artificial gravity, and allow for more radiation shielding of the craft. Nuclear rockets use

Top: **NASA's proposed uranium 235 powered Modular Bimodal Nuclear Thermal Rocket (NTR) Transfer Vehicle is designed to take astronauts quickly and safely to Mars and back.** *(Drawing courtesy of NASA's John Glenn Research Center)*

Right: **This SNAP-27 Radioisotope Thermoelectric Generator (RTG) (foreground, in center) was used by the crew of Apollo 14 to power science experiments on the moon. The mini nuclear power plant used plutonium 238 for its fuel source.** *(Photo courtesy of the National Aeronautical and Space Administration)*

lightweight hydrogen gas as their fuel source. As in an Earth-bound nuclear power plant that uses the heat of uranium fission to turn water to steam, a nuclear rocket uses this same heat to heat the hydrogen to provide thrust.

The many advantages of nuclear rockets will enable a larger spaceship to carry a larger payload to and from Mars in a much quicker timeframe than with conventional chemical rockets. Also, a chemical rocket would have to be many times more massive than the ones currently used to launch the space shuttle.

Below: **Nuclear power was also proposed to power an aircraft. This Convair NB-36H (a modified ten-engine B-36 intercontinental bomber) actually flew under conventional power with an operating nuclear reactor aboard.** *(Photo courtesy of the U.S. Air Force)*

NUCLEAR TREATIES

Between 1963 and 1996, the United States and the Union of Soviet Socialist Republics (USSR), now the Confederation of Independent States, and many other countries worldwide have been involved in ten different nuclear treaties and agreements. The first was the 1963 Limited Test Ban Treaty between the U.S., the USSR, and the United Kingdom. The most current treaty is the 1996 Comprehensive Test Ban Treaty (CTBT), an agreement signed by the same parties as the 1963 treaty, plus ninety other countries, that bans all nuclear testing. Other nuclear treaties have been signed to limit a country's maximum number of nuclear warheads and nuclear-capable missiles, and to try to prevent the spread of nuclear weapons to nonnuclear countries.

The U.S. conducted its last nuclear test on September 23, 1992. This less-than twenty-kiloton device was the 1,054th nuclear test by the U.S., and was similar in size to

American President John F. Kennedy signs the 1963 Limited Test Ban Treaty on October 7, 1963, banning nuclear testing in the atmosphere, outer space, and underwater. The treaty with the Soviet Union and the United Kingdom did not ban underground testing. *(Photo courtesy of the National Archives)*

the world's first nuclear test at New Mexico's Trinity Site on July 16, 1945.

OTHER USES

In addition to the use of nuclear materials for power production, in medicine, for timekeeping, and for weapons, another use of nuclear materials that is growing rapidly in the U.S. and across the world is the irradiation of food. Radiation is used to destroy most biological contaminants in certain foods, thus preserving the food for longer periods and solving some of the problems of storage and shipping, especially in developing countries.

Other new uses of radioactive materials include the sterilization of medical supplies, the conservation of water-damaged books (and other paper materials), and recently, the develoment by the International Atomic Energy Agency of nuclear techniques to detect and combat infectious disease, such as malaria, tuberculosis, and hepatitis.

The largest and one of the Museum's most historical exhibits is our Boeing B-52B. This 1962 photo shows "Old 0013" when she took part in Operation Dominic, a series of thirty nuclear tests that were conducted in the Christmas and Johnston Island areas of the Pacific. *(Photo courtesy of the U.S. Air Force)*

② About the Museum

A BRIEF HISTORY

In October 1999, the National Atomic Museum celebrated its 30th Anniversary. Founded as the Sandia Atomic Museum by the U.S. Army's Field Command, Defense Atomic Support Agency, the Museum has gone through numerous changes in management, mission, and physical appearance since opening on October 6, 1969.

The Museum's home since first opening has been Kirtland's Building 20358, which is leased from the U.S. Air Force. The building was originally built in 1951 as a heavy maintenance facility for the 90mm anti-aircraft guns that were stationed around Albuquerque and many other large American cities during the early years of the Cold War. Fortunately, the Soviet bombers for which they were waiting never arrived.

The creation of the Museum was the result of a six-year effort by a number of Sandia Base organizations (Sandia Base is now part of Kirtland AFB), and included the current operators of the Museum, Sandia National Laboratories. The effort

Built in 1946, Building 828 was the first building occupied by the Manhattan Project's Z-Division on Sandia Base in Albuquerque, New Mexico. The building was torn down in the spring of 1999 by the Z-Division's successor, Sandia National Laboratories. *(Photo courtesy of the U.S. Department of Energy)*

was started in late 1963 by the Commander, Field Command, Rear Admiral Emmet O'Beirne. The Admiral had visited the Los Alamos Science Hall and Museum (now the Bradbury Science Museum) shortly after it opened. He decided that a similar museum on Sandia Base would be an excellent community relations project where the long history of the Base and of nuclear weapons development could be told.

The 11:00 a.m. ribbon-cutting ceremony officially

The Museum's high bay in 1972, three years after the Museum opened. *(Photo courtesy of the National Atomic Museum)*

opening the Sandia Atomic Museum was officiated by New Mexico Lieutenant Governor E. Lee Francis and Major General John T. Honeycutt, Commander, Field Command, Defense Atomic Support Agency. Master of Ceremonies of the well-attended event was Sandia Base Commander, Colonel A. D. Pickard.

Four years later, in January 1973, the name of the Museum was changed to the National Atomic Museum to reflect its growing national and international audience, and the fact that it was the only public museum in the nation and in the world that preserved the history of nuclear weapons.

On March 17, 1976, under the threat of imminent closure by the military due to funding problems, the ownership/operation of the Museum was taken over by the Albuquerque Operations Office of the Energy Research and Development Administration (ERDA). ERDA was the predecessor of the current owner of the Museum, the Department of Energy (DOE). During the ERDA and DOE period of the late 70s and the 1980s the low bay area of the Museum was filled with exhibits on solar power and other alternative energy sources.

The Energy Reorganization Act of 1974 abolished the Atomic Energy Commission (AEC), successor to the Manhattan Project, and established the Energy Research & Development Administration (ERDA) and two other Federal agencies on October 11, 1974.

In late 1991, Congress granted the Museum a Congressional Charter as the official Atomic Museum of the United States. This was the result of the efforts of longtime Museum Director, Joni Hezlep, and others who had worked with U.S. Senators Pete Domenici and Jeff Bingaman and U.S. Congressman Steve Schiff (all of New Mexico). A result of this Charter was the creation of the National Atomic Museum Foundation. The Foundation operates the museum store, runs tours to important Atomic Age historical sites, and provides funding for Museum improvements, new exhibits, and many innovative educational programs.

In 1994, an extensive remodel greatly improved the exterior appearance of the Museum, added a fire-protection sprinkler system, and replaced most of the Museum's forty-year-old electrical wiring. The Museum's primary exhibit area, the high bay, was also remodeled.

On October 1, 1995, the day-to-day operation of the

After the operation of the Museum was taken over from the military by the Energy Research & Development Administration (ERDA) in early 1976, many of the new exhibits in the Museum reflected ERDA's focus at that time on conservation and alternative energy sources. *(Photo courtesy of Sandia National Laboratory)*

Museum was transferred from the DOE to Sandia National Laboratories. This transfer led directly to an almost half-million dollar renovation of the Museum's weapon exhibits (summer of 1996) and numerous other improvements in its staffing and operation.

A goal of Museum management has been to move the Museum off Kirtland Air Force Base to make it more readily accessible to the general public. A new, improved Museum, with twice the exhibit space will feature expanded exhibits, a larger theater, and many other enhancements that will enable it to better present the history, science, and future of the Atomic Age.

Facing page: **Ground Zero at New Mexico's Trinity Site National Historic Landmark, location of the world's first nuclear detonation on July 16, 1945, is marked by a small monument constructed with lava rocks.** *(Photo courtesy of the National Atomic Museum)*

Participants on a Museum tour to Yucca Mountain, Nevada, pose in front of the Tunnel-Boring Machine (TBM) that drilled a five-mile-long test tunnel into the proposed radioactive waste repository. The TBM can carve out a tunnel twenty-five feet in diameter. *(Photo courtesy of Steve Shook)*

OPPORTUNITIES AT THE MUSEUM

Education—The Museum's mission is to be the resource of nuclear history and science education. To attain this goal, the Museum offers an extensive education program directed at students of all ages, which are described in the Museum's annual Educator's Guide. Visit the Museum's WEB site at www.atomicmuseum.com for a copy of the Educator guide and for more information on all the following programs including the Museum's annual series of fun, week long Summer Science Camps for children ages eight to twelve. The exciting and educational "hands-on" ZOOMzone™, based on the popular PBS show ZOOM, is another great learning place for young children at the Museum.

Outreach—A new education program available from the Museum is our science and technology outreach van named *Up 'N' Atom*. The van will enable the Museum to take inter-active science and technology learning opportunities on the road to rural New Mexico children.

Receptions—The Museum can also be rented for receptions and special events, including sleepovers. Being

Three members of one of the Museum's popular week-long Summer Science Camps prepare to launch model rockets from Hardin Field, on Kirtland Air Force Base. *(Photo courtesy of the National Atomic Museum)*

surrounded by the powerful relics of the Atomic Age and our fascinating exhibits can make your event truly unique.

Tours—For children and adults, the Museum offers its Scientific Tour Series. In addition to the popular biannual tour to Trinity Site, the location of the world's first atomic test, Scientific Tours in recent years have visited the Nevada Test Site, Colorado's Cheyenne Mountain, and the Department of Energy's controversial WIPP (Waste Isolation Pilot Plant) Site southeast of Carlsbad, New Mexico.

Volunteer Programs—The Museum's unpaid staff is involved in virtually every aspect of the Museum's programs and operations. Volunteers as young as twelve years of age are matched with a wide variety of jobs based upon their skills and interests. Volunteering at the Museum is a great opportunity to meet new people, to learn new things, and to make a real difference in the lives of others.

KIRTLAND AIR FORCE BASE

The merger of Kirtland AFB and Sandia Base in 1971 resulted in a 51,558-acre base with over 21,000 civilian and military employees. While the Kirtland side (west) of the base can trace its history back to 1937 as Albuquerque's "new" municipal airport, the east side of the base (Sandia) dates urther back to May 1928 as the privately owned Albuquerque Airport.

Military activity at the municipal airport began in 1939 with the servicing of lend-lease military aircraft enroute to Great Britain in the early days of World War II. In January 1941, local efforts resulted in the construction of the Albuquerque Army Air Base (AAAB) adjacent to the municipal airport. By late 1941, pilot and combat crew training were well underway at AAAB, and during the subsequent war years the training of bombardiers, glider pilots, and B-24 crewmen filled the Base facilities. The Bombardier School was the Base's most important activity during the war, eventually graduating 5,058 U.S. Army and 81 Chinese Nationalist Bombardiers. In February 1942, AAAB was renamed Kirtland Field in honor of Colonel Roy C. Kirtland, an early military pioneer and one of the Wright Brothers' first military students.

To the east of Kirtland Field, the U.S. Army Air Forces had established a training center for aircraft mechanics and air depot service personnel at the old private Albuquerque Airport, now called Oxnard Field, on May 12, 1942. Renamed the Albuquerque Air Depot Training Station (also unofficially referred to as Sandia Base), this depot, upon the

completion of its original training mission, was used as a convalescent center for wounded aircrew under the name of the Army Air Forces Convalescent Center. When the Convalescent Center closed in April 1945, Sandia was used as a storage and dismantling facility for war-weary and surplus aircraft as the war finally came to an end in August 1945. Over 2,250 aircraft were melted down at Sandia, reclaiming some 10,000 pounds of aluminum, in addition to iron, copper, and other recyclable materials.

Both Sandia Base and Kirtland Field played important roles in the development of the atomic bombs by the Manhattan Project, serving as secure transportation, storage and supply facilities for airport-less Los Alamos, New Mexico, one hundred miles to the north. For a complete history of Kirtland AFB, visit <www.kirtland.af.mil>.

This late 1945 view toward the east and Tijeras Canyon shows some of the thousands of World War II aircraft stored on Albuquerque's Sandia Base waiting to be "recycled." *(Photo courtesy of the U.S. Air Force)*

③ Touring the Museum

LOBBY

Entering the Museum you first encounter the lobby desk, where friendly Museum volunteers will gladly answer your questions and give you an overview of the Museum and its contents. The lobby also contains a public phone, and is where you pay the Museum's low admission charge. Senior citizen, military, and youth discounts are available. A short hallway at the west end of the lobby provides access to the main section of the Museum and its many exciting exhibits.

On display in the lobby, is an excellent exhibit on Marie Sklodowska Curie and her husband Pierre Curie. Marie and Pierre were awarded the 1903 Nobel Prize in Physics (also shared with Antoine Henri Becquerel) for their original work on radioactivity. Five years after Pierre's untimely death in April 1906, Marie was awarded an unprecedented second Noble Prize in 1911 for isolating the highly radioactive metallic element radium. Marie Curie was the first woman to win a Nobel Prize and, so far, the only one to win two of the prestigious awards!

LOW BAY

The Museum's "low bay" contains a number of exhibits and provides access to the Museum Store, with its many unique items, the History Mystery Theatre, and the Museum's Research Library.

"Seeing is Healing," an informative exhibit on nuclear medicine, is one of the exhibits found in the low bay. Nuclear medicine involves the use of radioactive materials to help diagnose and treat a wide variety of diseases and disorders. Most people do not realize that in the United States alone, over 40,000 nuclear medicine prescriptions are written daily!

Adjacent to the nuclear medicine exhibit is the Museum's exciting ZOOMzone, which is based on "ZOOM!" the popular PBS children's show. The ZOOMzone encourages children to dream up their own science activities which they can share with other children via the ZOOM web site. ZOOMzones, found in children's and science museums across the country, are specially designed to give

Above: **"Marie Sklodowska Curie: Woman of Science," the Museum's informative exhibit on the Polish-born Marie Curie, highlights her life, work, and the winning of two Nobel Prizes for the groundbreaking scientific work conducted by Marie and her husband, Pierre, in the field of radiation.** *(Photo courtesy of Harold Rarrick)*

Facing page: **The Museum's colorful ZOOMzone is located at the south end of the Museum's low bay across from the Museum Store. ZOOMzone is based on the popular Public Broadcasting Station (PBS) children's show, "ZOOM!"** *(Photo courtesy of Harold Rarrick)*

families and school groups a creative and supportive environment to try hands-on activities they have seen on ZOOM. The Museum's ZOOMzone features computers for children to send their results and their own ideas into ZOOM Central where some of them may be incorporated into future ZOOM shows. The Museum's ZOOMzone also includes a mini-theater for viewing ZOOM episodes and workstations where creative young thinkers can work on puzzles, brainteasers, and other exciting challenges.

Opposite these two low bay exhibits is a temporary exhibit area and at its south end, the entrance to the Museum's Research Library. The Museum's Registrar and volunteers can assist you in your research on the Atomic Age. Computers are also available to aid you.

In the Museum Store, located at the north end of the low bay, members of the National Atomic Museum Foundation (NAMF) can receive a 10 percent discount on all purchases they make. The store can also be visited on line at <www.atomicmuseum.com>
(Photo courtesy of Harold Rarrick)

THEATER

The Museum's History Mystery Theater seats seventy-five and is equipped with three state-of-the-art projectors for video presentations. A small stage and theater lighting allows the Museum to present historical plays on various themes.

The Museum's feature film is David Wolper's *Ten Seconds That Shook the World*. This classic 1963 black and white production tells the story of the World War II Manhattan Project, and has been shown four or more times daily in the Museum's theater for over thirty years!

In addition to showing other films related to the history and science of the Atomic Age in the theater, the Museum presents the exciting multi-projector, interactive *Virtually Cool Tour*. The VCT highlights some of the innovative new technologies at Albuquerque's famous Sandia National Laboratories, which currently operates the Museum for the Department of Energy.

STORE

Operated by the nonprofit National Atomic Museum Foundation (NAMF), the Museum's Store contains many unusual and unique items for kids of all ages, including a collection of new and out-of-print books on the dramatic history of the Atomic Age. For that special child or grand-child, the store sells fascinating and imaginative toys, games, and videos. A variety of creative T-shirts and jewelry can also be purchased. The store can be found on the Museum's Web site.

The Museum Store is also the source of information and tickets for NAMF's "Scientific Tour Series" of Atomic Age sites in America's Southwest, including the 2,150-foot-deep Waste Isolation Pilot Plant (WIPP) near Carlsbad, New Mexico.

Below: **These replicas of Fat Man and Little Boy are located in Bay 3 of the Museum's high bay. The original Fat Man was a plu-tonium implosion weapon, while Little Boy was a much simpler and less powerful gun-type uranium 235 bomb.** *(Photo courtesy of Harold Rarrick)*

HIGH BAY

Appropriately named because of its ceiling height, the high bay is the Museum's largest exhibit space. The high bay contains the largest collection of unclassified nuclear weapons in the world that you can actually see and touch!

On entering the high bay from the low bay you encounter two exhibit timelines. One details the many significant scientific discoveries and events that led to the successful development of nuclear weapons during World War II by America's Manhattan Project. The other timeline is titled "Waging Peace: The Challenge of Nuclear Stewardship." It presents the history of arms control from the unpredictable war elephants of ancient times, to the deadly medieval crossbow, and up to today's nuclear weapons.

The Manhattan Project exhibit includes replicas of the Little Boy and Fat Man nuclear bombs that helped end World War II in August 1945. Both the Museum's Fat Man

The Museum's Boeing B-52B, No. 0-20013, arrived at Kirtland AFB from the Boeing factory in Seattle, Washington, on April 30, 1955. It left Kirtland for the last time sixteen years later in April 1971 by road, to the then Sandia Atomic Museum. *(Photo courtesy of the National Atomic Museum)*

Above: **The Museum's massive twenty-five-foot-long Mk 17 atomic bomb is similar to the one accidentally dropped just south of Albuquerque's airport by a ten-engine B-36J on May 22, 1957.** *(Photo courtesy of the National Atomic Museum)*

and Little Boy are painted to closely resemble the originals dropped on Japan.

The story of the Manhattan Project leads into the history of the development of numerous nuclear weapons and delivery systems during the Cold War, ending with the tearing down of the Berlin Wall and the breakup of the Soviet Union in 1989.

A long corridor on the east side of the high bay, the Atomic Gallery, provides a convenient return path from the

Originally exhibited inside the Museum's high bay, the Atomic Cannon was used by Sandia National Laboratories to develop nuclear capable artillery shells. Sandia donated the cannon to the Defense Atomic Support Agency (DASA) in July 1968, over a year before the Museum opened on October 5, 1969. *(Photo courtesy of the National Atomic Museum)*

high bay's north end, and is also used extensively to present temporary and traveling exhibits on various historical topics such as the Cuban Missile Crisis.

OUTSIDE EXHIBITS

The Museum's main outside exhibit area contains additional nuclear and thermonuclear weapons and delivery systems. The largest of these delivery systems is the multi-purpose vintage Boeing B-52B Stratofortress. Long the backbone of America's bomber fleet (since 1955), the last B-52s are scheduled to remain on active service in the Air Force until the year 2030 or beyond. The B-52 can drop conventional or nuclear weapons from altitudes in excess of 50,000 feet or

Facing page: **This Titan I Intercontinental Ballistic Missile (ICBM) photographed while being lowered into its silo, was the predecessor of the Museum's Titan II. While the Titan I carried the W-38 nuclear warhead, the Titan II carried the much more powerful W-53 warhead.** *(Photo courtesy of the National Archives)*

Located at the north end of the Museum's outside exhibit area, the Snark (from left), Mace, BOMARC, and Matador were the predecessors of today's much smaller and more powerful multipurpose cruise missiles. *(Photo courtesy of the National Atomic Museum)*

The Museum acquired its F-105D "Thud" when the fighter-bomber crashed on take off from the Albuquerque airport on June 12, 1981. Officially named the "Thunderchief," the veteran aircraft was en route to Davis-Monthan AFB, Arizona, from Maryland for storage. *(Photo courtesy of the National Atomic Museum)*

from only a few hundred feet at over 650 miles per hour.

Walk into the B-52's impressive bomb bay that could hold up to eighteen powerful nuclear weapons. Gently touch the massive 22-ton Mk 17 gravity bomb or imagine the 80-ton, 90-foot long Atomic Cannon racing down the highway at over 45 mph. On the west side of the Museum, you will find a 115-foot-long Titan II Intercontinental Ballistic Missile (ICBM), the largest silo-based missile built by the U.S., that could zoom toward its destination at 15,000 to 17,000 mph!

When exiting the front of the Museum, additional outdoor exhibits are found adjacent and to the south of the visitor parking lot. They include two U.S. Navy Terrier Missiles (directly in front of the Museum's entrance), a Vietnam-era F-105D fighter-bomber, and a World War II vintage four-engine Boeing B-29 Superfortress, the type of aircraft that dropped the atomic bombs on Japan during World War II.

The Atomic Age began at exactly
5:29:45 a.m. Mountain War Time on
July 16, 1945, with the Trinity Test.
The National Atomic Museum presents
the story of this test, once described
as the single most important event of the
twentieth century, along with
other history and science of this Age
that affects the future of our world.

Considered the most advanced heavy bomber of World War II, the fully pressurized Boeing B-29 Superfortress was armed with ten .50 caliber machine guns in four remotely controlled turrets and two additional guns in the manned tail position, for a total of twelve. *(Photo courtesy of National Atomic Museum)*